Lavoisier : the Father of the Modern Chemistry

Arthur Bower Griffiths
Lawrence J. Henderson
Charles F. McKenna

Lavoisier :
the Father of the Modern Chemistry

LM Publishers

I

Lavoisier : the great scientist[1]

At the time when so many countries were at war concerning the succession to the Austrian throne, there was born at Paris, on 26th August 1743, Lavoisier, who was destined in after years to revolutionize chemistry and ultimately to perish, or rather was murdered by the French communists, during the Reign of Terror which accompanied the first French Revolution. Our hero's father was a wealthy merchant, of a scientific bent of mind, and who encouraged the pursuit of knowledge in the son.

Lavoisier was educated at the College Mazarin, where he studied several sciences; and

---

[1] By Arthur Bower Griffiths

the knowledge gained in early years was invaluable for his future career. By dint of his genius there rose a *nouvelle chimie*: "La nouvelle chimie est toujours celle de Lavoisier." Lavoisier is, without doubt, the father of modern chemistry, as he overthrew Stahl's, or the phlogistic, theory; although every science is the accumulation of truths discovered by numerous workers in all ages.

The doctrines of Aristotle, and Byzantine, Egyptian, Arabian, and European writers concerning the "four elements"—earth, water, fire, and air—and other non-scientific theories were swept away by the master-mind of the great Frenchman. Even Priestley, Cavendish, Scheele, Macquer, and other distinguished men could not entirely divest themselves of the phlogistic theory; in fact, Black "was the only chemist of his age who completely and openly avowed his conversion to the new Lavoisierian doctrine of combustion."

In 1764 Lavoisier (having been called to the bar as an avocat) gained the prize awarded by the Government of Louis XV. for the best method of lighting the streets of Paris and other large towns, and elaborate experiments on the subject are to be found in the memoirs of the *Académie des Sciences*—"the greatest scientific body on earth"—as the illustrious Academy has been called by a distinguished American writer; and in 1768 Lavoisier was elected one of its members. Between that date and 1774 he published many papers on chemical, mathematical, and geological subjects—all of which are remarkable for showing the extraordinary ardour and devotion to science of this truly great philosopher. The accuracy of his work and reasoning powers will be found by referring to two of his papers in the memoirs of the Academy for 1770, in which he refutes the idea, held by many, that water could be converted into earth; and it is remarkable that in this his first *important* research he employs the balance—"the essential instrument of all chemical research." He heated water in a closed and weighed glass vessel for a hundred and one days, and at the termination of the experiment found that the vessel had lost 17·4 grains, and on evaporation of the water a solid residue weighing 20·4 grains was obtained—the excess being due to unavoidable experimental error. Lavoisier concluded that water when heated was not transmuted into earth, which was the theory entertained by the alchemists and some of the pneumatic chemists. He proved that the water dissolved some of the constituents of the glass—a conclusion confirmed by Scheele. This research had a far-reaching and an important bearing on the notions or theories of the times—theories that had existed for centuries were to be swept away by his clinching experimental proof of their absurdity; the art of alchemy and the pursuit of the philosopher's stone was rendered futile; but, above all, the experiments proved that the old alchemical idea of the transmutation to be false; and it led him to enunciate one of the most profound truths in science—the non-indestructibility of matter— *"rien ne se crée, rien ne se perde de la nature"* — matter is everlasting. In every chemical reaction there is no loss of matter, only new forms are produced. This is the basis of every chemical equation known at the present day; and, consequently, it is impossible to overestimate the importance of the *conservation of mass*, which Lavoisier established by means of the new instrument of precision—the balance; although it had been a philosophic theory of early Greek and Latin writers. The idea of the *mass* of matter was first shaped into an exact form by Galileo, and more especially by Newton, in the glorious age of the development of the principles of inductive reasoning enunciated by Bacon and Descartes in their philosophical treatises.

In 1772 Lavoisier published the results of his experiments on the calcination of metals, the burning of phosphorus and sulphur, and the increased weight in each case was due to absorption of air, and that when the calces of metals were heated with charcoal they were reduced to the metallic state. Twenty years before these experiments, and twenty-two years before the discovery of oxygen by Priestley, Voltaire came to the conclusion that the increased weight of iron after being heated in air was due to its absorption of something in the air: "*Il est très possible que cette augmentation de poids soit venue de la matière répandue dans l'atmosphère.*" (It is very possible that this increased weight is due to the material in the atmosphere.)

Although Priestley first isolated "dephlogisticated air" or oxygen, it was left for the genius of Lavoisier to first interpret the phenomenon of combustion; and in 1778 (the year that witnessed the deaths of Voltaire and Rousseau within thirty-three days of each other) he established the fact that oxygen was the universal oxidizing principle.

During 1775 and subsequent years Lavoisier expounded his views on the nature of respiration, fermentation, and combustion; and we have to this day excellent sepia drawings by Madame Lavoisier of her husband's experiments on respired air.

In 1783 the composition of water was discovered by Cavendish, and confirmed the same year by Lavoisier and Laplace. Lavoisier was then able to explain the reactions which take place when metals dissolve in acids; and when metals burn to form calces, that oxygen is instrumental in the combustion. From the work of Cavendish, he first clearly stated the compound nature of water and determined accurately its volumetric composition (*Kopp*). "Although Cavendish was the first to show that water is produced when the two gases hydrogen and oxygen are exploded together, it would yet appear that he did not fully grasp the fact that water is a *compound* of these two gases; it was left to Lavoisier to give a clear statement of this all-important fact, and thus to remove the last prop from under the now tottering but once stately edifice built by Stahl and his successors."

The *nouvelle chimie*, based on the theory of combustion or oxidation, was not accepted by the older chemists, and many distinguished workers remained followers of Stahl and his phlogistic doctrine, but unprejudiced and younger minds readily accepted the teachings of Lavoisier—certainly a revolution on the older views of chemical theory. This revolution brought upon him much odium and obloquy— so much so that his effigy was burnt at Berlin as a protest against his antiphlogistic doctrines; yet the same Prussian *savants* and students a few years later readily accepted the *nouvelle chimie*, and now his philosophy is the bulwark, the very foundation stone, of modern chemistry.

The following words of a well-known living scientist are applicable in Lavoisier's case: "It is extraordinary how slow people are to appreciate enthusiasm on the part of *real* workers; and it *is* a terrible crime to upset preconceived ideas. However, it is satisfactory to feel that one has helped to advance truth and upset falsehood, even should the effort prove painful to a certain section of fossilized people who are too lazy to work themselves, and who try to suppress the capacity of work in others."

How few think justly of the thinking few! How many never think who think they do.

In his *Traité de Chimie* Lavoisier described more fully the formation of oxides, and the phenomenon of combustion; and he proved (1777-78) that the calces of lead, tin, and mercury are oxides of these metals. But the phlogistic theory was still held by many honoured workers in chemistry, and they believed that in Cavendish's inflammable air (hydrogen) was the long-cherished, but undiscovered, phlogiston. Concerning the solution of metals in muriatic or vitriolic acid with the evolution of inflammable air, these phlogistonists stated that the metals lose phlogiston by the process, and that a calx is a metal minus phlogiston or "metallic spirit"!

The quantitative work of Lavoisier, his weighing and measuring, and the philosophical deductions of his experiments, completely shattered the theory of phlogiston—a theory which prevented the advance of science and proved to be a false doctrine, which unfortunately lasted nearly a hundred years. A theory is only useful when it explains the truths of science and helps the advancement of knowledge. Many theories have disappeared from the realms of science, and many are undergoing modification even at the present day.

About eleven years (1785) after the succession of Louis XVI to the throne of his grandfather, the phlogistic doctrine was completely overthrown—and this was entirely due to the researches, philosophical deductions, and writings of the great Frenchman.

Lavoisier gave the present accepted definition of an "element," although it may be remarked *en passant* that the researches of Lockyer, Ramsay, and others in the present day have somewhat modified our views concerning the nature of the chemical elements.

About the years 1785-87 Lavoisier, aided by De Morveau (a convert to the *nouvelle chimie*), Berthollet, and De Fourcroy, formulated a new system of chemical nomenclature which was greatly needed for the further development of the Lavoisierian chemistry: elements, compounds, oxides (peroxides and protoxides), acids, salts, etc., were defined, thus forming the basis of the chemistry of today. Lavoisier was the general who marshalled the isolated facts and coordinated them into one harmonious and stupendous whole. The work of the four French chemists and the teaching of Lavoisier were, by the close of the eighteenth century, universally accepted by chemists.

Lavoisier was the first to attempt the ultimate analysis of organic compounds, by burning these bodies in a stream of oxygen, and collecting the water and carbonic acid produced. This work is of historical interest as being the last performed by the illustrious savant, and just before execution.

But his ultimate right to fame rests on his interpretation of the phenomenon of combustion—that it is not a decomposition but a combination; the indestructibility of matter (conservation of mass); his philosophical deductions of the work of others; the introduction of the balance into all chemical operations; the recognition as elements of those bodies only which yield but one kind of matter; the formation of compounds from elements; his system of nomenclature; and, finally, his work (in conjunction with Laplace) on specific heats. Surely a goodly list, and sufficient to make his name immortal in the history of chemistry.

His energy knew no bounds, and at an early age, twenty-five, he became *fermier général* under the Government, and had the right, with others, of collecting the indirect taxes throughout France; a and this position during the Reign of Terror caused his downfall. He held the post of a *fermier général* until 20th March 1791, when the Assemblée Nationale suppressed the institution.

Lavoisier was wrongfully accused of mixing with the tobacco "water and other ingredients harmful to the health of the citizens"; and in May 1794—l'an deuxième de la République—he and twenty-seven of his colleagues were arrested; and on 2nd May Dupin (a member of the Government) brought the charges against the twenty-eight officials, and they were condemned to death.

The accused passed the night of 5th May in the dungeons of the Conciergerie, and were to be fed on black bread, but a generous friend, supposed to be Madame Lavoisier, secretly bribed the jailers, and provided the inmates of the prison with better food. This was no easy task, as the sentries were a double row of gendarmes one on foot, the other on horses and the passages were poorly illuminated by torches. This was the precaution taken to prevent the escape of the prisoners. Some were located in dungeons, where horrors of the most revolting kind had been perpetrated. Lavoisier and others were placed in the dungeon previously occupied by Marie Antoinette. The night was horrible. Only a few had obtained folding-beds, without mattresses or bedclothes, and others rested on the bare earth of the dungeons! On 7th May, on the eve of his death, Lavoisier wrote a letter to his cousin Augez de Villiers in which he said (in French):—

« J'ai parcouru une carrière assez longue, surtout très heureuse, et je crois que ma mémoire sera accompagnée de quelques regrets et peut-être de quelque gloire. Que pourrais-je désirer de plus?

Les événements dans lesquels je me trouve saisi m'éviteront, selon toute probabilité, les inconvénients de la vieillesse. Je mourrai dans toute la force de mon âge. C'est encore là un avantage que je dois ajouter à ceux dont j'ai joui.

Si je ressens quelques sentiments pénibles, c'est de n'avoir pas fait davantage pour ma famille, c'est d'être privé de tout et de ne pouvoir donner ni à elle ni à toi aucune preuve de mon affection et de ma reconnaissance.

Il est donc vrai que l'exercice de toutes les vertus sociales, que des services importants rendus à la patrie, une carrière employée utilement pour le progrès des arts et des connaissances humaines ne sont pas suffisants pour vous défendre d'une fin sinistre, pour vous empêcher de périr comme un coupable!

Je t'écris aujourd'hui, parce que demain il me sera peut-être pas permis de la faire, et parce que c'est une douce consolation pour moi de m'occuper de toi et des personnes qui me sont chères à ce dernier moment. N'oublie pas de me rappeler, au souvenir de ceux qui s'intéressent à moi et fais que cette lettre leur devienne commune à tous. Selon toute probabilité, c'est la dernière que je vous écrirai. »

Lavoisier had made an enemy of no less a person than Marat. In 1780, the latter wrote a "*détestable essai sur la nature du feu,*" and announced that it had received the approval of the Academy. Lavoisier flatly denied the assertion. This infuriated the hatred of Marat to such an extent that in his pamphlet, *L'Ami du Peuple*, written in 1791, he says: "*Je vous dénonce le coryphée, le charlatan Lavoisier, chimiste, fermier général, régisseur des poudres et des salpètres, qui a mis Paris en prison et a interrompu la circulation de l'air avec un mur...*"

De Fourcroy spoke indignantly of Lavoisier's sentence: "*L'homme qui aurait illustré son siècle par ses talents, qui aurait répandu ses lumières sur la société, dont les travaux auraient eu pour but d'instruire de rendre meilleurs et plus heureux les hommes, serait placé dans un même tombeau avec celui qui en aurait fait le tourment ou qui en aurait été la honté!*"

Yet De Fourcroy, De Morveau, Monge, and others who were his friends and pupils did nothing to save the head of their master. Jealousy—despicable jealousy—was the cause of these men forsaking their friend. "Sa supériorité," they confessed, stood in their way! They only talked, but did not act; and it is stated that De Fourcroy even contributed, by "*sa terrible accusation*," to the death of Lavoisier.

Petitions for the commutation of the death sentence were scornfully rejected. The Revolution knew no bounds—"away to the guillotine," "à la lanterne!" were the constant cries during the years of its existence. With tricoloured scarfs and Phrygian caps of crimson cloth they sang as they marched the streets of Paris that terrible song of the "Marseillaise":—

> *Allons, enfants de la patrie,*
> *Le jour de gloire est arrivé!*
> *Contre nous, de la tyrannic,*
> *L'étendard sanglant est levé?*
> *and the murderous levelling song "Ça Ira."*

Procès-verbal
d'exécution de
mort.

L'an *second* de la République Française, le *Dix* *huit Floreal* — à la requête du citoyen Accusateur public près le Tribunal Révolutionnaire, établi au Palais, à Paris, par la loi du 10 Mars 1793, sans aucun recours au Tribunal de cassation, lequel fait élection au Greffe dudit Tribunal séant au Palais, je me suis ——————————— Huissier-audiancier audit Tribunal, soussigné, transporté en la maison de Justice dudit Tribunal, pour l'exécution du Jugement rendu par le Tribunal *Aujourd'huy* —————— contre *Lavoisier* qui *le* —— condamne

à la peine de mort, pour les causes énoncées audit jugement, et de suite je l'ai remis à l'exécuteur des jugemens criminels, et à la Gendarmerie qui *l'ont* —— conduit sur la place de *Révolution* ou, sur un échafaud dressé sur ladite place, l ad *Lavoisier* , en notre présence, subi la peine de mort, et de tout ce que dessus ai fait et rédigé le présent procès-verbal, pour servir et valoir à qui de raison, dont acte.

Enregistré gratis, à Paris, le *8 Floreal* l'an deuxième de la République, une et indivisible.

Jugues

COPY OF THE DEATH WARRANT OF LAVOISIER

Terrible! terrible in those days! Men and women, tied down to a cart (tumbril), were hurried along the streets to the place of execution. Down clanked the axe, and the head of a victim rolled into the *corbeille*. The joy of the jealous, black calumnies, devilry, the hatred of cowards, the rage and stupidity of the masses, these were the feelings of men in the year 1794—and they triumphed.

Lavoisier asked for a short time to complete a research in which he was engaged, but Coffinhal (President of the Revolutionary Tribunal) remarked that *"la République n'a pas besoin de savants; il faut que la justice suivre son cours."*

On 8th May 1794, the immortal Lavoisier was guillotined in the fifty-first year of his age. Calm and resigned, he met his death without flinching, without demonstration, knowing that he had done his duty both to the State and to Science.

Lavoisier was President of the Académie, and a deputation of its members penetrated the prison and placed wreaths on his grave in the Conciergerie. The name of "Lavoisier" required no embellishment, nor was the sculptor's art needed to perpetuate it in posterity.

> No lengthen'd scroll, no praise-encumbered stone; □
> My epitaph shall be my name alone.*Byron.*

As long as chemistry exists the name of "Lavoisier" will always be remembered as the creator of a new era in science; and as long as the human race is capable of estimating the worth of noble labours, so long will the name of "Lavoisier" live, and the memory of him who bore it remain enshrined and held in affectionate reverence by succeeding generations.

Coffinhal, who would not listen to Lavoisier, had, says Lamartine, "the massive frame, figure, and masculine vigour of the Alpine races of his country (an Auvergnat). The energy of his mind responded to that of his muscles.

Payan was the head, and Coffinhal the hand, of this night and morrow." And of this man Robespierre said: "You destroy me, you destroy yourself, you destroy the Republic." Ultimately he was guillotined.

Concerning this period Carlyle says in his *French Revolution*: "The spring sends its green leaves and bright weather, bright May, brighter than ever; death pauses not. Lavoisier, famed chemist, shall die and not live. Chemist Lavoisier was Farmer-General Lavoisier too, and now all the Farmers-General are arrested; all shall give an account of their moneys and incomings, and die for putting water in the tobacco they sold. Lavoisier begged a fortnight more of life to finish some experiments, but the Republic does not need such; the axe must do its work."

To illustrate the feelings, animosities, jealousies, and fiendish propensities of Robespierre and his minions, two coins and a medal of the period bear ample testimony. The obverse side of one coin has the inscription: "Ludov XVI., D. Gratia" (with the head of Louis); and on the reverse side are the royal arms of France (fleur-de-lis) and the inscription: "1782, Rex Franciæ et Navarræ." The obverse side of the other coin, only ten years later, has the head of Louis and the inscription: "Louis XVI., Roi des Français"; and on the reverse side: "1792, 4 de la lib.," with the royal arms replaced by a floral design. There is a great difference in the title of the king; and the fourth year of liberty is most significant.

The medal is an English one struck in 1794, on the obverse of which is the following inscription: "A map of France, 1794. France divided, throne overthrown, honour trodden under foot, religion 'sixes and sevens,' glory erased, and fire in every corner." And on the reverse is the inscription: "May Great Britain ever remain the reverse."

These were the fanatical times in which Lavoisier lived and worked. Fancy any man of science in these days attempting to do research work during such a general upheaval.

A few words concerning the ever-memorable place of execution will be of interest. The Place de la Révolution, now the Place de la Concorde, the first square in Europe, has a tragic history. Originally a waste ground, it was reclaimed in 1748, after the peace of Aix-la-Chapelle (18th October 1748), and a statue of Louis XV. was erected there by the Municipal Council of Paris—the square then receiving the name of Place Louis XV. On 30th May 1770, at a display of fireworks to celebrate the marriage of the Dauphin, afterwards Louis XVI., with Marie Antoinette, a panic arose which resulted in the death of 1200 persons, and 2000 seriously injured. During the Reign of Terror in 1793 the guillotine was erected on the spot where now stands the Obelisk of Luxor (twin monolith to Cleopatra's Needle). Louis XVI. and Marie Antoinette were among the first victims, and between January 1793 and May 1795, upwards of 20,000 persons were guillotined. In 1799 the famous square was named the Place de la Concorde; it was afterwards renamed after Louis XV., and in 1826 after Louis XVI., and, finally, in 1830 it was again renamed the Place de la Concorde; and it was on the site where the Obelisk now stands that our hero Lavoisier was executed— his work unfinished, but he left a fine legacy to posterity.

His house was in the Place de la Madeleine, Paris; and in 1900 (a hundred and six years after his death) a beautiful bronze statue of him, by Barrias, which rests on a massive granite pedestal, was erected in the same square and opposite the house where Lavoisier lived. The pedestal bears the inscription: —

Antoine Laurent Lavoisier, 1743-1794, le fondateur de la chimie moderne. Erigé par souscription publique, sous le patronage de l'Académie des Sciences. M. Berthelot, Secrétaire perpétuel pour les sciences physiques.

*

Eleven years after the death of Lavoisier, his widow was married to Count Rumford—the founder of the Royal Institution, London. Two years after the death of Lavoisier, his wife published his *Mémoires de Chimie* (1796-99), in four volumes; and between the years 1857 and 1896, Lavoisier's *Œuvres Completes*, in six volumes, edited by Dumas and Grimaux, were published at the expense of the French Government.

*

We conclude our study of Lavoisier in the words of Galileo: "Ricordiamoci in grazia, che il cercar la constituzione del mondo è de' maggiori e de' più nobili problemi, che sieno in natura."

Lavoisier was nominated in 1768 to succeed Baron in the Academy of Sciences, by Lalande, who proposed him on the ground that he had knowledge, talent, and activity, and possessed a fortune, which, relieving him from the necessity of embracing another profession, would enable him to be very useful to science. His principal competitor was Jars, an eminent metallurgist. Lavoisier was chosen, but the final decision rested with the king, and his minister decided that Jars should have the seat. Out of deference to the views of the Academy, a new position of adjunct chemist was provisionally created for Lavoisier, with the understanding that on the occurrence of the next vacancy in chemistry he should go in without a new election. The vacancy occurred through the death of Jars in the next year.

Desiring, as the biographers pleasantly express it, to place himself on a financial footing in which he could pursue, independently, investigations involving costly expenditures, Lavoisier sought and obtained in 1768 a position as one of the farmers-general (of the revenue). He conscientiously performed the duties of his office; instituted reforms in taxation by removing useless duties, and earned the gratitude of the Jews of Metz by freeing them from an odious impost. M. Grimaux represents him as performing the duty of making regular tours of inspection, with which he associated the study of the features of scientific interest which the places he visited might afford. The work of this office brought him into association with farmer-general Paulze, whose daughter he married, and who went with him to the scaffold. In 1776 Turgot made him inspector-general of powder and saltpeter. In this capacity he made great improvements in the manufacture, so that, while he put a stop to forced official searches for saltpeter in the cellars of private houses, he quadrupled the product of the salt, and so increased the explosive force of gunpowder that the French brand became as much superior to the English as it had been inferior.

Lavoisier's great work consisted in the discovery of the true functions of oxygen and the nature of combustion; the determination of the relations of the solid, liquid, and gaseous states of matter; and in many other observations that embodied the germs of what have become since the leading principles of chemical science.

Oxygen was detected at about the same time by Priestley, Scheele, and Lavoisier; but the phlogistic theory of combustion possessed the minds of chemists, and, although Eck de

Suchbach and Jean Rey had already dimly discerned the truth, no one had paid any attention to their discoveries, and Lavoisier was working on what was to him, and substantially to the world, virgin ground. "Fixed air" and "combustible air" had been speculated upon, and "the air that is left after combustion" had attracted attention. But the phenomena of this kind, inconsistent as they were with the phlogistic theory, had not been sufficient to overthrow it.

The first germ of Lavoisier's theory on these matters was embodied in a sealed packet which he deposited with the Academy in 1770. Recognizing that the calcination of metals could not take place without the access of air, and that the freer the access the more rapid the calcination, he "began to suspect," as he expresses himself, that some elastic fluid contained in the air was susceptible, under many circumstances, of fixing itself and combining with metals, and that to the addition of that substance were due calcination and the increase in weight of metals converted into calxes. From this thought came, after much groping with erroneous conclusions, the idea that air is a compound containing a vital part and another part, and that it is the vital part that is absorbed. The behavior of charcoal when burning in oxygen pointed to the nature of that substance and to the true theory of combustion. This new vital substance, which, uniting with metals, formed calxes, and with other substances generated acids, he called *oxygen* or the acid-producer; the air that was left after combustion was *azote,* or lifeless. The inflammable air which, combining with oxygen, was found to form water, was called *hydrogen.*

Upon these facts, and with a few other names of known substances, Lavoisier constructed the system of chemical nomenclature which, after having undergone many modifications to conform to new discoveries, still rules. The "muriatic radicle" gave Lavoisier some trouble, for he could find no oxygen in muriatic acid, and his experiments upon it with oxygen resulted in the production of a neutral substance which must be its calx; and so he called chlorine oxidized muriatic acid. Such mistakes were natural in the early days of chemistry. The decomposition of volatile alkali, or ammonia, by Berthollet, led to the suggestion which Lavoisier gave out with great modesty, that many earths, still regarded as simple, might be compound; and that their apparent indifference to oxygen should be attributed to their being already saturated with it.

On the nature of gases and vapors, which had not been understood before, Lavoisier asserted, in a memoir published in 1777, that most bodies were capable of existing in three different states—those of solids, liquids, and vapors, or aeriform fluids. The terms airs, vapors, and aeriform fluids express only a single form of matter—a class of bodies infinitely extended; and this principle "gives the key to nearly all the phenomena relative to the different kinds of air and to vaporization." While heat tends to change volatile bodies into vapor, the pressure of the air has a contrary effect; and "the tendency of volatile bodies to evaporate is in direct ratio to the heat to which they are exposed, and inverse to the weight or pressure brought to bear upon them." Lavoisier's memoirs on heat, expansion and contraction under changes of temperature, and latent heat, show an insight into the accepted principles. He discussed with much sagacity the question whether heat is a fluid or a force; and it would not be hard, for one who is determined to look for it, to find in his essays on this subject a prevision of the current constitutional chemistry. Lavoisier's later labors were physiological. They include papers on the production of carbonic acid in respiration and the office of the lungs in the process, in which the present theory is proposed as a secondary hypothesis, and on cutaneous transpiration. In his physiological studies, M. Dumas has found that he had arrived at a remarkable anticipation of modern views concerning the relations of organic to inorganic nature.

Lavoisier carried his energy into several other fields, and made his mark in all. He cultivated an estate of two hundred and forty arpents in the Vendôme, and in nine years doubled its production. His name is associated with a number of propositions looking to the public welfare or economical reform. In 1789 he presented in the National Assembly a report of the "Caisse d'Escompte," to which he had been attached for one year. As commissioner of the treasury he proposed in 1789 a new plan for the collection of imposts, which he elaborated in a special essay entitled "The Territorial Wealth of the Kingdom of France," a work which, according to M. F. Hoefer, in the "Biographie générale," gave him a place in the front rank of the economists of his time. He participated in the work of the commission on a new system of weights and measures. As treasurer of the Academy he set the accounts and inventories in order, and discovered some forgotten funds of the institution, and made them available. "In short, Lavoisier was to be found everywhere; and his facility and zeal, equally admirable, were adequate for everything."

On the 2d of May, 1794, twenty-eight of the farmers-general, of whom Lavoisier was fourth on the list, were accused in the Convention of conspiring with the enemies and against the people of France. On the 6th of May they were all condemned to death, and on the 8th were executed together. Lavoisier and his friends hoped that his great scientific eminence and the undoubted useful character of his career might be brought to bear to save him. Some efforts were made to exert such influence. Lavoisier himself drew up a memoir of what he had done for the Revolution. The Bureau of Consultations presented a detailed report on his labors. A deputation of the *Lycée des Arts* visited him at the *Conciergerie,* bearing "to

Lavoisier, the most illustrious of its members," a testimonial of its admiration.

Lavoisier left no children. He is described as having had a pleasing, intellectual face, and having been of large figure and of pleasant, sociable, and obliging disposition.

II

Lavoisier and the Rise of Modern Chemistry[2]

Lavoisier's achievement consisted in his recognition of the fact that weight is neither increased nor diminished in chemical changes, and in the elevation of this discovery, which has since been many times confirmed with ever-increasing accuracy, into the guiding principle of chemical investigation, the law of conservation of mass. This advance involved the introduction of the balance as the chief instrument of chemical research. Lavoisier's great success depended, further, upon the fact that he chose the process of oxidation and reduction (the reverse of the reaction of

---

[2] Lawrence Joseph Henderson

oxidation) for study. Not only is oxygen the most active of chemical elements, if both intensity and variety of chemical behavior be considered, and far the commonest upon the earth's surface, but also the most important chemical processes are reactions of oxygen.

The partial tearing off of oxygen from the carbon of carbonic acid and the hydrogen of water is the first step in the formation of all organic substances in the plant, and the recombination of oxygen with plant products the chief chemical activity of the animal. All this and much more Lavoisier recognized, and thereby revealed the true nature of another great phenomenon of nature. These investigations also disclosed, in the sequel, the chief source of all the energy which is available for the purposes of man.

It is only the energy stored up in the plant (originally the energy of the sunlight shining upon the green leaf of the plant and transformed by the action of chlorophyll) which is contained in all coal, wood, all kinds of oil, including petroleum, alcohol, in short every fuel. And it is exclusively by the union of the fuels with oxygen once more to form water and carbonic acid that this energy is liberated, as in the human body itself, and utilized by man. The resulting water and carbonic acid can then be used over again by the plant. The nature of this cycle of matter was clearly recognized by Lavoisier. This is the basis of nearly all our industry and commerce.

Conclusion[3]

The most advanced chemical philosophers of his day taught that there was something in every combustible substance which was driven out by the burning, that the reduction of an oxide of a metal to the metallic state meant the absorption of this substance or principle, which Stahl had called phlogiston. Lavoisier studied the teaching of the phlogistonists, but having also a mastery of physics and of pneumatic experimentation he became dissatisfied with their theory. He seized upon two important discoveries, that of oxygen by Priestley (1774), and that of the compound nature of water by Cavendish (1781) and by a masterly stroke of genius reconciled discordant appearances and threw the light of day upon every phase of the

---

[3] *By Charles Francis McKenna*

world's reacting elements. His theory, for a long time thereafter known as the antiphlogists' theory, was really the reverse of that of the phlogistonists, and was simply that something ponderable was absorbed when combustion took place; that it was obtained from the surrounding air; that the increase in the weight of a metallic substance when burned was equal to the decrease in the weight of the air used; that most substances thus burning were converted into acids, or metals into metallic oxides. Priestly had called this absorbed substance or gas dephlogisticated air; Scheele called it empyreal air; Lavoisier "air strictly pure" or "very respirable air" as distinct from the other and non-respirable constituent of the atmosphere. Later, he called it oxygen because it was acid-making (*oxys*, and *geinomai*).

So great a change ensued in experimental chemistry, and in theory and nomenclature, and such a mass of facts was coordinated and explained by Lavoisier that he has been justly called "the father of modern chemistry." He was the first to explain definitely, the formation of acids and salts, to enunciate the principle of conservation as set forth by chemical equations, to develop quantitative analysis, gas analysis, and calorimetry, and to create a consistent system of chemical nomenclature. He made deep researches in organic chemistry, and studied the metabolism of organic compounds. His memoirs and contributions to the Academiy were of extraordinary number and variety. His life in other fields was romantic, full of interest and a social triumph, but sadly destined to end in tragedy. Happily married, and having the aid of his wife even to the extent of employing her in the prosecution and recording of his experiments, he drew around his fireside and to his library at the State Gunpowder Works a circle of brilliant French savants and distinguished travellers from other lands. Early in his career he felt the need of increasing his resources to meet the necessities caused by his scientific experiments. With this in view he became a deputy *fermier-général*, whereby his income was much increased. But joining this association of State-protected tax-collectors only prepared the way for many years of bitter attack and a share of the public odium attaching to their privilege. He headed many public commissions requiring scientific investigation, he aimed at bringing France to such a state of agricultural and industrial expansion that the peasant and the working-man would have profitable employment and the small landed proprietor relief from the burdensome taxes hitherto purposely increased to make grants to corrupt favourites of the Court. Having incurred the hatred of Marat he found himself, together with his fellow *fermiers-général*, growing more and more unpopular during the terrible days of the Revolution. Finally in 1794 he was imprisoned with twenty-seven others. A farcical trial speedily followed, in which he was charged with "incivism" in that he had damaged public health by adding water to tobacco. He and his companions, amongst them Jacques Alexis Paulze, his father-in-law, were condemned to death. Lavoisier, who was devotedly attached to him, was obliged to stand and see M. Paulze's head fall under the guillotine, 8 May, 1794. Lavoisier was then 51 years old. His biographers say little as to his last hours.

LAVOISIER MONUMENT.
ERECTED IN PARIS BY INTERNATIONAL SUBSCRIPTION.